Ramos Rodriguez

THE PROMENADE OF ORDOÑO II

Ramos Rodriguez

THE PROMENADE OF ORDOÑO II

A HISTORICAL TOUR THROUGH THE URBAN PLANNING AND ARCHITECTURE OF THIS AREA.

ScienciaScripts

Imprint
Any brand names and product names mentioned in this book are subject to trademark, brand or patent protection and are trademarks or registered trademarks of their respective holders. The use of brand names, product names, common names, trade names, product descriptions etc. even without a particular marking in this work is in no way to be construed to mean that such names may be regarded as unrestricted in respect of trademark and brand protection legislation and could thus be used by anyone.

Cover image: www.ingimage.com

This book is a translation from the original published under ISBN 978-613-9-44116-7.

Publisher:
Sciencia Scripts
is a trademark of
Dodo Books Indian Ocean Ltd. and OmniScriptum S.R.L publishing group

120 High Road, East Finchley, London, N2 9ED, United Kingdom
Str. Armeneasca 28/1, office 1, Chisinau MD-2012, Republic of Moldova, Europe
Printed at: see last page
ISBN: 978-620-8-21351-0

THE PASEO DE ORDOÑO I I

A HISTORICAL OVERVIEW OF THE URBAN PLANNING AND ARCHITECTURE OF THIS AREA

ADRIÁN RAMOS RODRÍGUEZ

INDEX

1-INTRODUCTION

1.1. SUMMARY AND KEYWORDS

This paper aims to review the bibliography that analyses the space currently occupied by the Avenida de Ordoño II in the city of León. This is one of the main communication routes located in the centre of this medium-sized city in the northwest of Spain. The sources consulted study this space in its historical development from its beginnings and origins as an urban space or road, to the present day, converted into one of the communication arteries for people and cars, but also for the commercial, economic, administrative and residential flows of this city.

For this reason, this work will analyse this road space in its development and chronological evolution. This will be done in an interdisciplinary way, trying to take into account the evolution of all its social, cultural, economic, infrastructural and historical aspects, of course. We will also try to analyse the concepts implicit in the analysis of this question, such as the notions of urban centre, product, merchandise or instrument. It will also include the current state of this space within the city of León and its possible changes in the near future.

urban centre, historical evolution, city of León, urban expansion, market

1.2. CONCEPTUAL INTRODUCTION

Avenida Ordoño II, in the capital of León, is the city's main thoroughfare, running partially through it from east to west. In its last stage, before its extension in 2011, the avenue was half a kilometre long and 30 metres wide. It ran between the squares of Santo Domingo and Guzmán. In 2011, on the occasion of the construction of the new railway station, it was more than doubled in length. The avenue crosses the Bernesga River and reaches the junction with Avenida del doctor Fleming .[1]

In this paper I propose to analyse the Avenida de Ordoño II as one of the most decisive parts of the last century, from the urbanistic point of view of our city. To do so, and in accordance with the study carried out by María del Pilar Durany Castrillo, "*La calle Ordoño II de León; De calzada real a eje comercial*", we can better understand the profound changes in this artery of the city which began its first steps in the final years of the 19th century.

Avenida Ordoño II, as an active "living" urban element, has been growing and modifying its structures gradually and over dozens of years. From what was originally the so-called Paseo de las Negrillas, to its current appearance as a lively commercial street and main artery of the city. It is largely due to what was its main urban impulse, the project for the widening of the city from 1897 onwards. It was finally approved in 1904, based on factors as decisive as economic, social and spatial factors, which turned it into a "collector of floating dynamism"[2] .

For this author, there is a gradual positive growth in the city towards the new city street, which contains an ever-increasing volume of business and traffic. Durany speaks of three different vital rhythms in this avenue: "morning or work

[1] Mª del P. DURANY, *La calle Ordoño II de León: De calzada real a eje comercial y de servicios*, Salamanca, 1990, p. 17.

[2] Ibid, *Op.cit*, p.18.

and traffic, at midday; the commercial bustle, in the evening; and the calm and quiet, at night". To better understand what this avenue was like in the middle years of the last century, the author quotes a popular saying: "Calle Ordoño abajo, los hombres lo ganan.//Calle Ordoño arriba, las mujeres lo gastan.//Calle Ordoño adentro, los bancos lo guardan" (Ordoño Street below, men earn it.//Calle Ordoño Street above, women spend it.//Calle Ordoño Street inside, the banks keep it)[3] .

From its development as a main road, Ordoño II evolved from a quasi-residential area to a residential and commercial area. The latter being the one that is mainly capturing the street's main raison d'être. There is a progressive outsourcing of a space in the city that serves as a focus of commercial attraction and services. For the city, the transformation of Paseo de las Negrillas into Avenida de Ordoño II is one of the best examples of the process of economic and urban development in the city of León .[4]

---

[3] Mª del P. DURANY, *Op.cit*, p.19.

[4] Ibid, *Op.cit*, p.20.

2-CONCEPT OF THE CITY CENTRE: PRODUCT, INSTRUMENT AND COMMODITY

The three concepts that condition the urban centre are: the product, the instrument and the commodity.

2.1-Town centre

We approach this artery by coining the term "**urban centre**". Durany (1990) states that two significant concepts fall within it: we are talking about a geographical place and we are talking about its social content. This is not the utopian case of that ideal city centre of the Renaissance, of Sforzinda, where the palace, the gardens and the privileged redoubts delineated in that urban epicentre are imagined. Here we must speak of an urban centre conditioned by the reality that influences the way it is executed and developed . [5]

2.2-Product

The urban centre can therefore be understood as a product, the result of the relationships between the elements that make up an urban structure. Without forgetting that it has two elements intrinsic to its meaning: content and form. We say that its **sociological expression** translates into content (**the living part of the urban subject**, its residents, its commercial and economic flows, its visitors and passers-by...). And on the other hand, like any living ecosystem, it has a spatial reference, which translates into form and materialises in the successive historical configurations .[6]

[5] Mª del P. DURANY, *Op.cit*, p.20.
[6] Ibid, p.21.

Space has a purely formal aspect, where we find its architectural and town-planning outlines, which will almost always be determined by social relations and the economic determinism that may be established there.

2.3-Instrument

Space, apart from reaching us and being understood as a product, according to Durany (1990) is used in parallel as an instrument, and as such is linked to those processes of exchange that emerge in its context. The management apparatus, whether municipal or private (we will see later that they sometimes coincide), will regulate and project the space according to profitability criteria rather than social reasons.

2.4-Commodities

Throughout the 19th century and much of the 20th century, in some periods more than others, the concept of land as a commodity was a constant. The arrival of the rural population in the cities brought with it the need to produce space, housing, to house all this population. Cities grew and the value of land multiplied. Land appeared as one of the most profitable investments for the wealthy social class. Rural land, associated during the 19th century with the Paseo de las Negrillas, was understood from the 20th century onwards, under the pseudonym of "Calle Ordoño II", as pure merchandise with which to satisfy the growing commercial demands that were appearing at that time .[7]

---

[7] Mª del P. DURANY, *Op.cit*, p. 21.

3- AGENTS INTERVENING IN URBAN SPACE. CONTRIBUTORS TO CHANGE

In order for urban renewal action to be set in motion, a number of agents must intervene and act as dynamisers of change. Although the process of urban transformation is marked by fixed patterns, it will vary according to the type of owners who own the means of production and the regulatory framework in which they operate.

The owner of urban land will try to obtain the maximum profit from the sale of his plot of land, while the property developer wants to buy the goods at a low price and then benefit from the profits he will make depending on the location of the building. As we will see later on, urban centre space achieves the highest prices within a city.

The improvement in the exchange value of a property implies a revaluation in the use value of a plot of land that can from now on be nurtured by the property developer. A plot of land in the desolate Paseo de las Negrillas is not the same as the same space in the current "Paseo de Ordoño II". In this way any improvement introduced in the land can provide a potential valuation in its exchange value .[8]

But how is this preponderance of exchange value built over the initial use value, at what point are new needs subscribed that did not exist in the original layout, since when did it begin to be identified as the urban centre of our city and who were these agents who intervened in it, how did they build the modern signifiers that we find today when we pass through, no longer remembering those unrecognisable humble and peripheral origins, how did they build the modern

[8] Mª del P. DURANY, *Op.cit*, p.22.

signifiers that we find today when we pass through, no longer remembering those humble and peripheral origins?

These are all questions that require us to search for the reasons that will provide us with answers. As we have seen, there are many conceptual outlines and we must delve into them. And we need to know the interests that facilitated the gentrification of the place and the projects that shaped its outlines and why.

4-WHEN DEVELOPMENT BEGINS. HISTORICAL CONTEXT OF URBAN DEVELOPMENT.

All this has to do with what happened at the beginning of the last century and with the appearance and approval of the so-called "ensanches" of the city. The urban renewal or extension operations appeared as a solution to remedy the crisis or deterioration of the central areas[9] , and also to meet other needs.

The first phase of the generalised process of industrialisation, which became evident in Spain at the beginning of the 19th century, led to an explosion of the traditional morphology, surprised the city and destroyed the established structures, affecting, for better or worse, its urbanisation. The decomposition of agrarian social structures forced emigration to urban centres. This displacement also coincided with an enormous increase in population, as a large part of it went to the cities[10] . This was mainly reflected in the old town centres, where the proletarian class was crammed together.

In our case we are talking about an elitist leisure area that does not want to share these precarious conditions and needs its own space and urban demarcation. On 9 April 1842, when the national law was passed declaring "the right to freely lease urban estates with the conditions they wished to stipulate (...)", the liberation of urban property from feudal burdens was consummated, thus accommodating it to the political and economic demands of the still young liberal regime.[11]

In Spain, the industrial revolution arrived later and the urban changes it brought about were of a different order to the international panorama. Old town centres were not always devastated. Spanish urban planning law was clearly

---

[9] Ibid.

[10] Mª J. GONZÁLEZ ORDOVÁS, *Políticas y estrategias urbanas*, Madrid, 2000, pp.121-122.

[11] Ibid. p.63-65.

expansionist and focused on the expansion of towns as an alternative to interior remodelling. [12]

Although the city predates industrialisation, both industrialisation and chaotic urban growth radically transformed the landscape that captured the rational attention of bourgeois society. After setting the objectives, choosing the means and weighing the consequences, bourgeois society initiated a new form of social action: town planning and urban planning, and inaugurated urbanism as a technique and an ideology used for the benefit of the economically more powerful classes. Urban planning is part of the planning effort, it is a project of action and represents a rehearsal for mastering the future, an effort of prospection and spatial orientation.

City planning as a conscious attempt to encompass and coordinate all aspects of an urban ensemble is characteristic of bourgeois society, for although political power had previously had an impact on spatial form and visual symbolism, its pretension had not reached the stubborn bourgeois aspiration to comprehend and exhaust all effects.

The generalised discourse at the end of the 18th century on the power of space and the space of power precedes the elaboration and deployment in the 19th century of a whole series of strategies aimed at dominating and manipulating space with a view to maintaining and securing their interests and capacity to act, or to questioning and modifying it in others. A decisive link in the process which, linked together, forms part of the course of the construction of the Modern State and its rational and geometric layouts.[13]

---

[12] Mª del P. DURANY, *Op.cit.* p.20.

[13] Mª J. GONZÁLEZ ORDOVÁS, *Op.cit*, pp.129-130

Returning to "our ensanche", it does not follow the example of Madrid, and is more similar to the Barcelona project. The fact that the ensanches were configured or planned as spaces that could be assimilated to the old city meant that they were immersed in the historical process of urban transformation operations. Thus, today, the urban centre is identified in some cities with the old quarter, while in others the ensanche has become the functional centre of the city, such as Calle Ordoño II. [14]

Simplifying the consequences, and taking this context to our local terrain, we can say that there are now two parallel effects on the urban physiognomy. At the same time as working class families were settling in the old town and suburbs, there was a flight (ordered and planned) of the bourgeois classes towards the Ensanche, which had appeared as an urban solution to the saturation of the old town. We have seen how the cities became catalysts for the rural exodus linked to industrialisation and how they had to take on a greater number of "citizens" from an urban planning point of view.

[14] Mª del P. DURANY, *Op.cit,* pp.20.

5- THE CONFIGURATION OF A DYNAMIC URBAN SPACE.

Normally these urban renewal operations are carried out on already produced, consolidated spaces, but as we have already seen, this is not the case in León. The following historical events are of enormous importance here: the confiscation of ecclesiastical property, the arrival of the railway in the city and, more generally, the rationalist conception that 19th century men had of space, which was significantly reflected in the alignment and expansion plans. These three interrelated facts, together with the increase in population at the end of the last century, resulted in a shift of the bourgeois social class towards the widening of the city.[15]

The reason why this urban space later attracted commercial activities, until then located in the old quarter, is in line with the development of the changes mentioned so far. These three sections that we will now see are the identifying pillars of its morphology, and without their progression in the form of a sum, it would have been difficult to find the Calle Ordoño that today channels us.

5.1-DISENTAILMENT

Although the first disentailments began in the reign of Charles III, the disentailment process can be considered a fundamental event from 1798, with the reign of Charles IV, from Godoy to Mendizábal, passing through Cádiz[16] . They consisted of the nationalisation of the lands or assets of the "dead hands", belonging to ecclesiastics or civilians. The second and third stages, which include the work of Mendizábal with the Royal Decree of 19 February 1836 and the Law of 29 July 1837 (with its complement by Espartero in September 1841), are considered to be those which aided this process of renewal, including Madoz's

---

[15] Mª del P. DURANY, *Op.cit.* pp.24-25.

[16] Ibid. p.25.

Law of 1 May, the latter being considered as a third phase of the Spanish disentailment process which mainly affected municipal property.

The objectives were: to meet the needs of the public finances and to transform the "legal regime of agrarian property", which was scored as a triumph of the bourgeois revolution. Property went from being concentrated in the hands of the clergy to being in the hands of a wealthy bourgeoisie. It was a simple transfer of ownership, giving rise to land speculation, which ceased to be cultivated and became part of the urban building land. Once this social class had acquired ownership of the land, it created, together with the public authorities, the legal regulations that favoured its interests .[17]

The formerly rustic (and later urban) property between the current boundaries of Calle Ordoño II, i.e. from Sto. Domingo to Glorieta de Guzmán, were all confiscated lands. The parent properties existing at the time of the confiscation belonged to the following ecclesiastical communities: Hospital de San Antonio Abad, Cabildo de la Catedral, Colegiata de San Isidoro and Fábrica de la Iglesia del Mercado. The date when another property was vacated is not known, and everything seems to indicate that it belonged to the Convent of Santo Domingo, to which it adjoined.

As soon as the clergy's property was auctioned off, large landowners began to appear in the area. The disentailed estates had excellent agricultural resources, as they had water at the foot of the land. Although the valuation of these areas in terms of their size in rural terms can be considered normal, this valuation must be made taking into account their location. At the time of the disentailment, these were farms leased by the local people. After this, the day labourers' contracts were cancelled and this peri-urban agricultural area soon

[17] Ibid. pp25-26.

became urban land where buildings, initially few in number, gradually replaced the haystacks and orchards .[18]

Would the widening and alignment of streets and the renovation of the hamlets have been possible if the ownership of the land had remained in those dead hands? For professors Tomás Cortizo and Antonio Reguera, if the disentailment had not happened, this would have been a serious obstacle to urban renewal and the future expansion and growth of the city .[19]

5.2 - THE ARRIVAL OF THE RAILWAY

Fig. 1. Image of the first railway station in the city of León, which was inaugurated in 1863, although this image was taken in 1882.

[18] Ibid. pp26-27.

[19] R. ARCE BAYÓN, *La ciudad de León en el siglo XIX. Transformaciones Urbanas Precursoras del plan de ensanche,* León, 2012, p.120.

Note: Image adapted from Olaizola Elordi, J. [Juanjo] (18 December 2023). The railway arrives to León [blog entry]. In: Historias del Tren. https://historiastren.blogspot.com/2023/12/el-ferrocarril-llega-leon-i.html

The arrival of the railway in León, under the "general railway plan" in the mid-19th century, was the main driving force for the development and economic revitalisation of the city. The state was the main promoter of this initiative. The development of this new communications network in Spain was not so much a response to a real need for new channels of communication and distribution, since production was scarce and did not require new markets to be communicated at that time, but rather a response to planning with a view to favouring the effects it could have on the industrial development that was already appearing in the second half of the 19th century.

In the case of León, apart from the aforementioned, there are other reasons why the capital of León was one of the chosen nuclei as part of the railway network. It is essential to mention Ignacio Gómez de Salazar, engineer of the same, who in 1855 presented to the Provincial Council a series of considerations about the importance of a railway through León, whose central axis is the exaltation of man as a producer being who should develop the "love for work".

He proposed a route linking the capital with the centres of mining production in the north of Spain, which would be complemented by the agricultural production of Castile. In addition to promoting the exchange of goods, it would enable new forms of capitalist production to enter the traditional structures of production. All this would have a decisive influence on the shaping and development of cities, with the railway being considered one of the main driving forces of urban growth, especially in León .[20]

[20] Ibid. pp.185-188.

Ordoño II Street, also known in the past as Calzada de Santo Domingo, Calzada Real, Calzada del Príncipe Alfonso, Carretera del Estado... acquires its true relevance from this fundamental fact: the arrival of the railway. This means of transport made its entrance in León in 1863, not without great technical and economic problems.

The execution of the railway project dates back to the end of the 19th century and it is the most important and transcendental objective for the development of the economic and social interests of León. The institutions of the capital, City Council, Provincial Council, Government, etc., fought tenaciously to achieve it. In addition to donating assets from their own wealth, an appeal was made to the city's major taxpayers to buy shares in the railway in order to alleviate the high costs and to be able to achieve the project.

Some authors, including D.C., argue that the push for the arrival of the railway was "one more means for the wealthy class of the city to expand their business". It was not, he says, in the national or social interest". However, as the subject of this paper does not deal with this issue but with the consequences it had on the urban structure of the city, we will stick to it .[21]

The location of the station is no coincidence and will have a great impact on our main subject, the Avenida Ordoño II. The options that were considered were located on both banks of the river. The Town Council owned land in Quiñón de Vega (located on the right bank of the Bernesga), land that in 1860 was ceded to the Board of Agriculture under the pretext of being able to use part of it (without compensation) if the railway project required this space. And so it would be done. The station was finally located on the right bank of the river, after the publication of the Royal Order of 1861[22] . Over the years, the station would serve as a pole

[21] Mª del P. DURANY, *Op.cit.* pp.27-28.
[22] *Ibid.* pp. 28-29.

of attraction for the expansion of the city, an expansion controlled and directed by the bourgeoisie through the expansion project.

The quickest and easiest way for the population to access the station was the Negrillas road, which soon became the main street of the urban expansion. The railway left the necessary and sufficient space for the expansion of the city to develop between the railway and the old quarter, channelled by means of an urban plan that favoured it, especially for certain owners who needed this beneficial situation. Calle de Ordoño II thus acquires an important function: that of being a link and communication route, and with the railway.

5.3 - THE ALIENATION AND ENLARGEMENT PLANS.

At the beginning of 1863, three main roads were proposed to connect the town with the railway: the Renueva, the central or Santo Domingo and San Francisco. At that time, the importance of regularising these paths was made clear, giving them sufficient width for future needs.

Fig. 2. First plans of the Ensanche of the city of León in the 19th century.

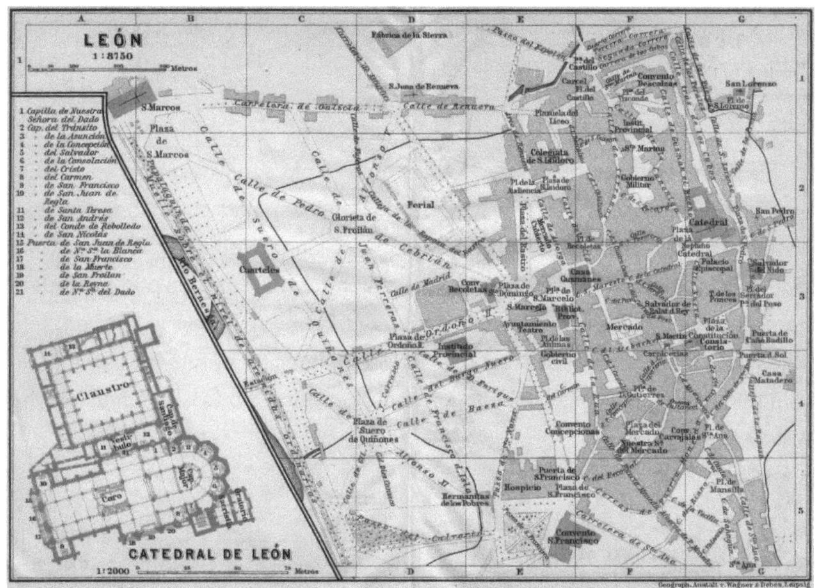

Note: Image adapted from Unkown (14 January 2013). El Ensanche de León [Blog entry]. In: Walking BCN. 2012-13. https://caminarbcn12-13t.blogspot.com/2013/01/el-ensanche-de-leon.html

The street alignment plans involve new layouts, which implies a redefinition of plots and the extension of the network of building plots, "which on the other hand allows a mutual relationship between the circulatory system and the building system"[23] This model is worked here within the rationalist conception of space of the end of the 19th century. Through this, a greater optical-aesthetic effect is achieved and the space is hierarchised by means of categories applied to the streets. The widest streets will be of the first category and order, and the rest of the streets will be classified in decreasing order, according to the municipal ordinances. These systems become more complex as the socio-spatial relations of the city intensify. This is complemented by another series of rules that regulate the height of buildings, the number of floors, etc.... And which are no more than

[23] *Ibid.*

the normative systematisation of the use of a basic space that is going to have highly profitable real estate effects.

The first project for the enlargement of the city of León was drawn up in 1897 and its report consists of three parts: a geographical and political description, including a justification of the chosen area; a second part dealing with the living conditions of the city, materials used, dimensions...; and finally, a part devoted to the provisions adopted for the enlargement of León. This plan is justified as a necessity with two main causes: the unaffordable costs of remodelling the old town and the population's desire to expand westwards, up to where it meets the railway.

The original layout of the widening projected four oblique streets in Ordoño II. One that, crossing the Gran Vía, would lead to Calle la Torre (leading to San Isidoro). Two more: one leading to Gran Vía and another from Avenida la Independencia. And another that led to what is now República Argentina. This layout caused unease among some property owners as, among other things, the oblique layout of the branches caused all the plots of land bordering and adjoining neighbouring properties to lose value and the triangular surfaces that were produced to be lost. Finally, by means of a letter to the town council, several property owners subscribe to this complaint and propose a model of transversal and parallel streets to that of Ordoño II. In this way, they create smaller blocks that extend more metres of frontage and produce better-used plots. This is how it was finally approved by the Consistory.

Fig. 3. Plan of the Ensanche at the beginning of the 20th century.

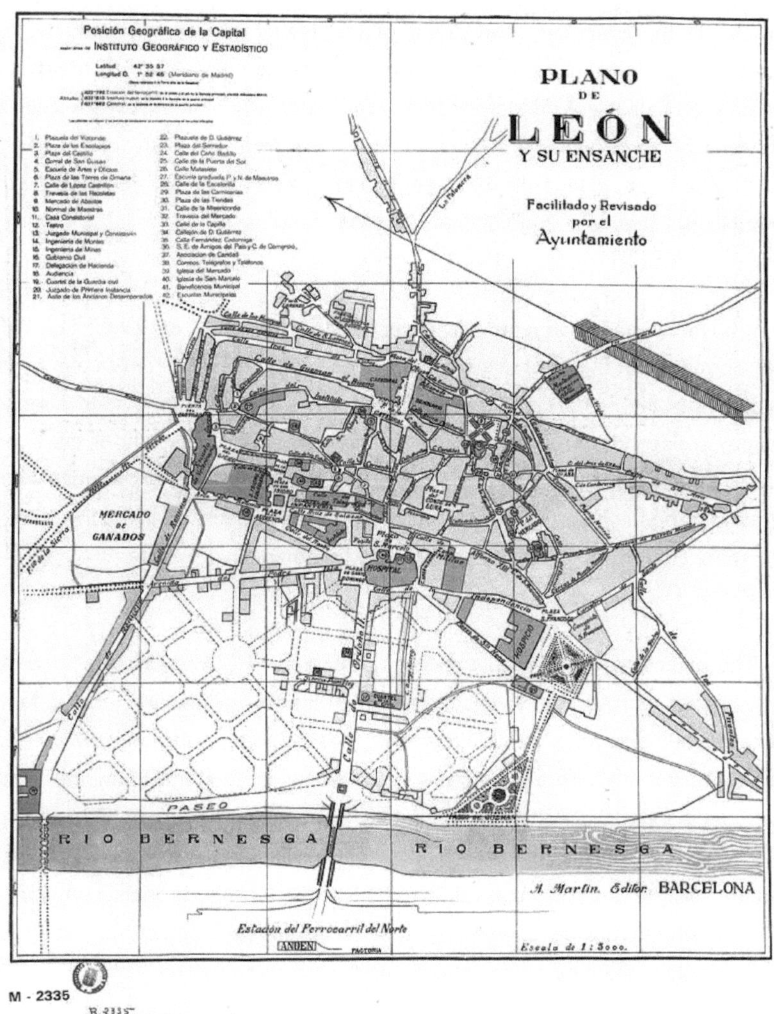

Image adapted from Unkown (14 January 2013). El Ensanche de León [Blog entry]. In: Walking BCN. 2012-13. https://caminarbcn12-13t.blogspot.com/2013/01/el-ensanche-de-leon.html

The reform project proposed in 1905 and approved in 1907 introduced variants in accordance with these claims, the oblique routes were eliminated in almost all the branches, and some of the "small squares" created by the chamfered junction of these perpendiculars were also discarded in order not to lose the profitability of their properties. If the widening of the city was initially an unnecessary and overly ambitious project, now, from the first decade of this century onwards, it began to make some sense, as it acted as a receptor for the population, initiating a stage of intensification of the buildings .[24]

The dynamic agents involved in the process of expansion and transformation of the urban structure have been represented in this case by the demands of the owners of Calle Ordoño II. They did not allow "their" street to occupy a secondary role in the city. In this way, the profitability of the space delimited by the main axis of the widening will be extended in accordance with a process of opening up new roads, which give greater accessibility to this axis, at the same time as extending the profitability to the adjoining space.

The extension of the road system is promoted by private owners, who see their capital improve in this way due to the profitability of the land, both in terms of its use and exchange value. If a property has frontage on Ordoño it will be very profitable, but it will be even more so if it extends its frontage to another street such as Alfonso V. For this reason, the vast majority give up part of their properties for these openings, and some, although this should be the responsibility of the town council, even take charge of the urbanisation work of the new streets (paving, sewage...). Calle de Alfonso V is one of the first streets to be legally opened by means of a dossier in November 1914.

The most intense process of street opening began around 1915. Prior to this date there were only two: Sierra Pambley and Travesía de D. Cayo, today "Alcázar de Toledo" and "Capitán Cortés", opened to the public in 1908 and 1939

[24] Mª del P. DURANY, *Op.cit.* pp.31-38.

respectively. Since the last reform of the widening plan in 1935, the Ordoño II axis has had seven streets and two perpendicular passages. It is significant that the widening did not consolidate, neither urbanistically nor demographically, until after the first quarter of the last century. This indicates how unnecessary the first widening plan was at that time. The owners did not start to build their "chalets" until the arrival of the railway, with the most intense period of occupation and building starting in 1915.

Fig. 4. Chalet de Don Paco built in 1894

Note: Image adapted from Olaizola Elordi, J. [Juanjo] (18 December 2023). The railway arrives to León [blog entry]. In: Historias del Tren. https://historiastren.blogspot.com/2023/12/el-ferrocarril-llega-leon-i.html

The social aims, an argument defended by many of the promoters of this idea of urban expansion, were never the driving force behind this initiative. Rather, it was a pretext for materially developing an area of the city from which they could extract important economic benefits[25] . We shall now see if the occupation of the site has anything to contribute to this urban development.

5.4 - THE ENSANCHE AND SURROUNDINGS DURING THE POST-EARTHWAR period

The designation of the city of León as provincial capital (1833), the arrival of the railway (1863) and the confiscation of land at the end of the 18th and beginning of the 19th century, boosted the urbanisation of the land where the Ensanche leonés is located today[26] . But also, the rural exodus - increased in the post-war period - and the exponential growth of its inhabitants - from 22,000 in 1920 to almost 50,000 in the 1940s - favoured the development of this new urban area and its growth in the following years . [27]

The Ensanche project - like the reforms undertaken at the end of the 19th century in the old quarter - is understood partly as a political-bourgeois response to the problems derived from the demographic growth[28] , but also as a residential and speculative opportunity in which some wealthy citizens invested in order to settle outside the walled enclosure[29] . What is interesting for our study is that it meant the possibility of creating a new city epicentre; more prosperous, healthy and ordered than the old[30] , which during the first third of the 20th century

[25] *Ibid.* pp.38-41.
[26] Mª del P. DURANY, *La calle Ordoño II de León: De calzada real a eje comercial y de servicios,* Salamanca, 1990, pp. 24-27.
[27] VV. AA., *León, Casco Antiguo y Ensanche. Guía de Arquitectura*, León, 2000, pp. 23-31.
[28] Mª J. GONZÁLEZ ORDOVÁS, *Políticas y estrategias urbanas*, Madrid, 2000, pp. 121-122.
[29] Mª del P. DURANY, *Op.cit.,* pp. 23-38.
[30] J. R. CABALLERO CHICA, *La arquitectura de la ciudad de León en su fase inicial (1907-1919),* Master's thesis defended at the University of León, León, 2017, pp. 28-30.

materially strengthened its expectations[31] and which in the post-war period was consolidated and began to mature as such .[32]

This Ensanche was born at the end of the 19th century as a genuine proposal in our community, but it was not until after the post-war period that the spatial definition of its layout[33] was completed, nor the occupation of most of its plots[34] . By then, although it had evolved according to the interests of its owners, its limits were already similar to the current[35] . It was bounded to the south by the old winter promenade - today Calle Lancia - and to the north by the old Convent of San Marcos. In the centre was the "Paseo de las Negrillas" -today Ordoño II-, which was the backbone and the link between the railway station and the historic city, and the latter, together with the river Bernesga, marked the east and west limits of the Ensanche . [36]

Thus, in architectural-urban terms, the Ensanche was a real revolution, since at the dawn of the 20th century there were more than 70 hectares of free land where, unlike the historic centre, modernity would have a place[37] . As a result, new forms, layouts, techniques and materials, typical of the great capitals, appeared in it[38] . And as for the historicist styles of the first buildings[39] , it is striking that they were similar in appearance both to those built decades earlier on the

[31] S. TOMÉ FERNANDEZ, "La segunda fase de ocupación del Ensanche Leonés: el proceso de renovación desde los años 60", in L. LÓPEZ TRIGAL (Ed.), *Los Ensanches en el urbanismo español, el caso de León,* Madrid, 1999, pp. 115-119.

[32]T. CORTIZO ÁLVAREZ, "El Ensanche de León. Project and first occupation", in *Ibidem,* pp. 104-112.

[33] E J. R. CABALLERO CHICA, *Op.cit.*, pp. 28-30; T. CORTIZO ÁLVAREZ, *Op.cit.,* pp. 104-112.

[34] S. TOMÉ FERNÁNDEZ, *León, los ríos en el paisaje Urbano,* Gijón, 1997, pp.59-60.

[35] Mª del P. DURANY, *Op. cit.,* pp. 24-38.

[36] VV.AA., *León, Casco Antiguo y Ensanche...* p. 31.

[37]. J. R. CABALLERO CHICA, *Op.cit.*, pp. 117-118.

[38] *Ibid,* pp. 168-176.

[39] M. SERRANO LASO, *Arquitectura doméstica en León a principios de Siglo (1900-1923). La pervivencia del eclecticismo*, León, 1992, pp. 25-50.

alignments of the old quarter and to those built fifty years later under Franco's dictatorship . [40]

This parallelism was due to the fact that they were either the same architects or the continuators of the previous ones, and, in any case, all of them had received an academic and historicist training in Madrid to later practice their profession in cities such as León . [41]

---

[40] J. HERNANDO CARRASCO and M. SERRANO LASO, "Arquitectura contemporánea. Del neoclasicismo a la posmodernidad", in *Historia del Arte en León*, chap.16, León, 1990, pp. 263-264.
[41] M. SERRANO LASO, *Op.cit.,* pp. 117-118.

6-OCCUPATION AND DENSIFICATION OF SPACE IN THE 20TH CENTURY

The resident population of the capital of León at the beginning of the 20th century was around 16,000 inhabitants and the slow growth of the city was being perfectly absorbed by the old quarter and the suburbs. In 1910, it had a total of 1770 buildings, 82.4% of which were built inside the city walls. The rest were outside this area, which includes the widening of the city and therefore Calle Ordoño II. Even so, it should be pointed out that the population that moved and settled in the Ensanche and, mainly in Ordoño II, was a commercial bourgeoisie, which slowly and gradually brought about spatial changes and transformations. The type of building constructed was the "hotelito" or "chalets" of one or two floors, although four-storey buildings also began to be built, all of them luxury residences for the time. In 1923, with the demolition of the Hospital de San Antonio Abad, the Calle Ancha-Ordoño II axis was formed, as the land in the Plaza de Santo Domingo, which had previously been occupied by the walls of this enormous hospital building, became available .[42]

6.1-Land ownership

One of the variables that determines the process of urban renewal is the degree of concentration of land ownership. In the middle of the last century (1844) there was an important change in rural and urban property. As we have already seen, the disentailment actions led to the nationalisation of ecclesiastical property, mainly. When they were put up for public auction by the State, the highest bidders gained access to the property, who were undoubtedly certain wealthy families (both local and foreign), who saw the confiscation process as a good opportunity to invest their capital.

An example of this in our city was Cayo Balbuena and his wife Asunción Aguirre, who acquired and accumulated around 50,000 $m.^2$ in estates that formed part of what would later become the Ensanche. One of these "meadow-gardens",

---

[42] *Ibid.* pp.40-41.

measuring 1,393.54 m^2 , which formed part of the Hospital de San Antonio Abad, was acquired in 1856 by Asunción Iriarte together with another from the same estate measuring 1,567.7 m^2 , corresponding to the houses from numbers 1 to 11, both included, and had their boundary in what is now Burgo Nuevo .[43]

6.2-Process of urban space formation

The buyers were either engaged in the liberal professions or in commerce, so that it was this fact that marked the patterns of spatial growth, as it should not be forgotten that the owners were the main dynamic agents shaping the urban structure. They tried to make the most of their investments in rustic property, selling it as urban property at a higher price.

The disentailed estates or those belonging to private individuals were divided into numerous plots which were sold to other private individuals; the segregation and aggregation of plots was a constant in the process of the formation of the urban space to which they contributed a certain dynamism. In the case of the Balbuena family, in 1900 they owned 21.6% of the plots facing Ordoño II and by 1930 they only owned 8.1% of them.

These bourgeois families became related to each other, which led to a concentration of land in a few families, but numerous hands. This meant that either the heirs sold to a single owner on the premise of an indivisible estate or by common agreement for greater profit. In any case, the more heirs, the more division of land. This, (exclusively in a place as sought-after as the city centre) is an important added value. Little by little, the urban space was being configured, coinciding with the beginning of the building process, at the same time as the opening of the new streets and their urbanisation .[44]

---

[43] *Ibid.*

[44] *Ibid.* pp.45-48.

6.3-The building: the process of construction of Calle Ordoño II

The first glimpses of construction were made in the middle of the 19th century. When private individuals gained access to the confiscated properties, the building process began, which intensified during this century until it reached its most complex stage today. The first assessment of the new urban space, Calle Ordoño II, came from the public authorities themselves, who, in the Memoria del plan de Ensanche, stated that the street had "superb buildings that are the best in the whole city". Before 1890, there were only four buildings, with ten in 1900 and fifteen in 1904. When the ecclesiastical properties were confiscated, there were only two haystacks on the Negrillas road with ground and first floors . [45]

Fig. 5. Photo of the Paseo de las Negrillas at the beginning of the 20th century.

Note: Image adapted from Vergara Pedreira, S. [Susana] (4 April 2021). Those four houses of Ordoño. La Revista de El Diario de León.

[45] *Ibid.* p. 48.

https://www.diariodeleon.es/monograficos/revista/210404/1251483/cuatro-casas-ordono.html

The first buildings were erected on the grounds of Cayo Balbuena. In 1904, the number of existing buildings was 40.5% of what it is today, and the first four- and five-storey houses (including attics and ground floors) were already being built. By 1918, the number of houses had risen to 29, which was already 78.3%. And in 1930, 94.5% of the 37 plots of land that make up the street today were occupied; it is this last stage that coincides with the highest degree of occupation of the Ensanche by the population. Between 1890 and 1904 there was the highest rate of increase in the number of buildings per plot. But from 1918 onwards, the increase in occupation and densification took place in terms of height, from 6.6% of five-storey buildings in 1904 to 31.4% in 1930. These figures already show us at this time how highly valued land was as a commodity in Calle Ordoño II .[46]

6.4-Land uses

Initially, the Ensanche was configured as an eminently residential area, only accessible to wealthy families. But other land uses will gradually be introduced. New functions that complement or replace residential use. These are offices and shops that will gradually modify the socio-economic structure of the street.

The location of certain organisations in Calle Ordoño II favoured its promotion as a focus of attraction. In 1918, three important offices of the Public Administration were located in this area: the Civil Government, a department of the Town Hall (house no. 12) and the offices of the Tax Office (house no. 17). There was also a bank, the Banco Mercantil (in house no. 2). The commercial function was still scarce, but its later development was already beginning to be glimpsed. There was a pharmacy (house nº4), a hat shop (house nº2), a Singer's

[46] *Ibid.* pp. 48-49

shop (nº4), a tobacconist (nº23), a photography shop (nº7) and four food and beverage shops that began to supply the basic needs of the population living in the street .[47]

By 1936-37 there were already twelve shops selling basic necessities, thirteen premises dedicated to the hotel and catering trade, another thirteen textile shops, two perfumeries, eight agencies and offices in Ordoño, so the street specialised in tertiary uses, with only two factories, one for soap and another for pasta and chocolate at this time. The drifts are towards food shops, fabrics and the hotel and catering trade, without losing the residential and communication uses towards the station .[48]

According to the information provided by the 1958 Tax Contribution, of the 37 houses that make up the street, only six (16.2%) are used only as residences, although there were no buildings completely dedicated to tertiary functions, taking into account the number of premises per building dedicated to this use. The process of occupation and densification of the urban space of Ordoño II was almost completely consecrated at the beginning of the 1960s.[49]

The building of the old Town Hall of León was completely refurbished and enlarged in December 1962, under a project by Prudencio Barrenechea Sánchez, for which the neighbouring infrastructures of the Teatro Principal[50] and the Gota de Leche[51] were demolished. Unlike the 1940 proposal, the aim was not to dismantle the 16th century façade, but to recover the well-preserved parts and recreate the most deteriorated ones in style, adding some new elements and

[47] *Ibid.* pp. 51-53.
[48] *Ibid.* pp. 51-53.
[49] *Ibid.* pp. 51-53.
[50] E. FERNÁNDEZ GARCÍA, *León y su actividad escénica en la segunda mitad del siglo XIX*, Doctoral thesis defended at the UNED, Madrid, 1997, pp. 70-78.
[51] S. SANTOS VALERA, "La gota de leche en la ciudad de león: una institución benéfica municipal", *Argutorio*, nº10, 2003, pp.27-28.

modifying the interior of the building to make it more functional[52] . It was more preservationist than the post-war work, and also functional and regenerative-symbolic, similar to the work carried out at the same time on the town halls of other cities .[53]

The building that housed the Civil Government, now the Subdelegation of the Spanish Government in León, was inaugurated on 28 June 1947 by the Minister of the Interior[54] . Its previous location - in the 1930s - may also have been the Ensanche area, next to the former Provincial Treasury Offices[55] . Another idea is that, before the Civil War, it was located on the site previously occupied by the eclectic and historicist style factory, Zarauza, between Fajeros, Héroes Leoneses, Padre Isla and Gran Vía de San Marcos streets . [56]

In any case, on 9 January 1940, the transfer of the land was approved for its construction in the current site[57] , known today as Plaza de la Inmaculada, which during the dictatorship was christened Plaza de "Calvo Sotelo"[58] . This shows us that during the Franco regime, squares or central places in towns continued to be chosen as the ideal places to locate the representative bodies of the regime and public life[59] . And so, on 22 September 1943, the relevant expropriations were approved[60] and on 30 March 1944 the first stone was laid . [61]

---

[52] VV.AA., *León, Casco Antiguo y Ensanche...,* pp. 96-97.

[53] M. ANDRÉS EGUIBURU, *Op.cit.,* pp. 113-116.

[54] VV.AA., "Crónica de cien años 1901-2000", *El siglo de león*, León, 2001, p. 277.

[55] T. CORTIZO ÁLVAREZ, *Op.cit.,* pp. 104-112.

[56] J. R. CABALLERO CHICA, *Op.cit.*, pp. 123-124; J. C. PONGA MAYO, *León perdido...*, pp. 130-131.

[57] VV.AA., "Crónica de cien años...", p. 249.

[58] J. C. PONGA MAYO. *El ensanche de la ciudad de León...,* p. 196.

[59] M. ANDRÉS EGUIBURU, *Op.cit.,* pp. 110-113.

[60] VV.AA., "Crónica de cien años...", p. 232.

[61] *Ibid*, p. 249.

6.5-Demographic dynamics and structures

Urban space is, above all, a social product. Therefore, the study and analysis of the socio-professional category will be one of the variables that will mark the guiding thread of the urban renewal process. As we have seen in the dynamisation of other structures, such as buildings, its configuration will go hand in hand with the population occupation of calle Ordoño II, which grows slowly and gradually. Thus, in 1897, the Negrillas street had 228 inhabitants. In 1917 the street already had 429 inhabitants and, thirteen years later, in 1930, 709 people lived there. In 1960 it reached its peak of 1011 inhabitants. As far as the demographic structure is concerned, the population pyramids of Calle Ordoño II up to 1960, in line with the development of the street itself, show a young population, in contrast to the old quarter, which has an ageing demographic structure .[62]

[62] *Ibid.* pp. 53-55.

7-CHANGES IN LAND USE AND OCCUPATION IN THE MID-TWENTIETH CENTURY: THE TERTIARISATION OF ORDOÑO II

As for the city of León, as in other cities, urban speculation was not corrected either, as land continued to be commercialised, as had been the case since before the war[63] . And both the east and west pavements of the Bernega river[64] , as well as the neighbourhoods of San Mamés, San Esteban or Las Ventas to the north, grew speculatively[65] . Even the neighbourhood of San Claudio, which was next to the Ensanche, grew in a much more disorderly way than the latter, which shows the lack of interest in caring for, ordering and cleaning up urban developments outside the city centre[66] .

Calle Ordoño II underwent a series of transformations from the 1960s onwards and it is now when this spatial framework acquires, like the rest of the Ensanche, a significance as a "capital accumulator". The largest labour supply in the city is represented by tertiary activities, administration, transport, commerce and banking. The liberalisation of the economy led to an increase in purchasing power at this time and, at the same time, the development of a large number of state-supported property developments, thus promoting the organisation of an increasingly strong property market.

The demand for land is organised on the basis of spatial location, and developers become part of the body of dynamic urban agents, as they transform the spatial reality according to the profitability of the land, new land divisions emerge according to the new land use and a densification of space takes place .[67]

---

[63] as a vital framework and a collective phenomenon, and the fragmentation and private appropriation of it". See: A. T. REGUERA RODRÍGUEZ, "Especulaciones urbanísticas en el León de posguerra", *Tierras de León*, nº 68, León, 1987, pp. 3-9.

[64] S. TOMÉ FERNÁNDEZ, *León, los ríos en el paisaje...*, pp. 59-61.

[65] V.V.A.A., *León, Casco Antiguo y Ensanche...*, pp. 34-35.

[66] J. C. PONGA MAYO. *El ensanche de la ciudad de León...*, pp. 36-37.

[67] *Ibid.* pp. 55-56.

From the moment the bourgeoisie settled in this area, it began to be considered a highly valued place, but it was its location as an urban centre, its accessibility, the densification of the population in the surrounding area and, of course, as we said, the capital accumulated in this place that really provided the use and exchange value that we now associate with Ordoño II .[68]

7.1-The transition from vertical to horizontal ownership

Although some properties still belong today to those bourgeois families who became owners at the beginning of the 20th century, the vast majority now belong to the entities and private owners of housing and commercial premises and bank offices, these new strategically located agents have given a new functionality to the street.

According to Professor López Trigal, from 1973 onwards, there was a period of banking expansion that has come to be known as the "bank boom". The increase in the number of bank branches has its origin in the increase in customer deposit accounts". The location of these banks, created from 1975 onwards, was based on the line of concentration of the previous branches in the city: a "city" or headquarters of bank offices in the commercial centre of the city, Santo Domingo and its radial streets.

Around 1980, four more banks, together with the Banco de España, joined the ranks of the new property owners. Only three of them used the entire building for their own purposes. The others reconstructed the buildings and divided them horizontally, selling the newly established flats to private owners.

[68] *Ibid.* pp. 55-56.

In 1960 the Regulatory Norms of Horizontal Property were established (Law of 21 July 1960) and with these general provisions a new system of land ownership was introduced whereby a building would no longer have one owner, but the number of owners would vary according to the number of flats created and the investment capacity of the applicant. The leasehold system used until the 1960s was in decline. As horizontal division has been applied to the buildings, the tenants who occupied the houses have had access to private property and although it is a voluntary decision, it implies the abandonment of the property by the tenant so that another potential buyer can gain access to it .[69]

This phenomenon of land subdivision has been exploited by the small commercial bourgeoisie of León and by the management and administration companies of the urban spatial centre for a better and more efficient development of the exchange of their products. The relationship between land prices and rents according to the benefits of its central location is positive for the investor .[70]

7.2 The renovation of buildings.

The 1976 Law on the Regime of Land and Urban Planning includes some articles alluding to the promotion of building, whose objectives are to accelerate the renovation process, for example, those "estates on which there are paralysed, ruinous, demolished or unsuitable constructions for the place on which they are located..." are considered plots of land and it is made a condition to build within two years, after processing and registration in the register of municipal plots of land.

The public administration favours the renovation of buildings. Under this last category of "inadequate" and following the profitability of the land, it was logical

[69] *Ibid.* pp. 57.
[70] *Ibid.* pp. 57-60.

that the space should be renovated in accordance with the new uses and functions of the land. The new buildings tend to rise in height, with lower spaces for shops and offices, the rest of the space being used for residential purposes. Most of the buildings we can see on the street today are 5 or more storeys high. Even so, of the 37 buildings with a façade on Ordoño, only 9 have been demolished and rebuilt.[71]

7.3-The renewal of land use since 1960.

From the 1960s-70s onwards, there was an explosion in prices coinciding with the liberalisation of the Spanish economy and the increase in the standard of living. The exchange value is increasingly higher and therefore the use value increases, which affects the functions of the space, which is becoming more and more tertiary. The food sector is losing representation in parallel with the detriment of the residential character of the street.

The renovation of buildings promoted by the new land uses and the new identity of the street has mainly affected the south pavement (odd numbers), but the renovation of uses has also been carried out on the north pavement (even numbers), with activities being adapted to the old housing structures. It is also significant that the greatest concentration of activities and renovation of buildings is concentrated in the area closest to the Plaza de Santo Domingo, the nerve centre of the city.[72]

---

[71] *Ibid.* pp. 61-62.

[72] *Ibid.* pp.63-66.

7.4-Demographic analysis from 1960 onwards.

The analysis of the social fabric that resided in the area, both the socio-professional category and the residents of the area, is fundamental to understanding the transformations of the street from the 1960s onwards in comparison with its beginnings.

The population has been declining since the 1960s due to population displacement to other areas and a declining birth rate, as well as the ageing of older residents.

It is still a bourgeois, upper-middle class population and this can also be seen in the socio-professional category that has settled in the area: lawyers, doctors, engineers, architects, financiers have acquired property to develop their profession, and not so much as a residence.

The process of degradation of the old quarters has slowed down as a result of the heritage and tourist interest they hold, but it is no match for the care given to these expansion areas. We conclude that the demographic contingent, which gives a social content to the geographical space, is closely related to the process of transformation affecting the area.[73]

---

[73] *Ibid.* pp.66-69.

8-CONCLUSIONS

I have decided to make an "alternative" conclusion, I confess that I set out to develop a work that was too ambitious for my limited experience in urban planning. I have seen this throughout the elaboration process, and when I read the conclusions of the works I have used as references, I find incredibly greater riches than I could bring to the table.

For this reason I have chosen a more accessible option for this last part. A small formal conclusion, more evaluative than conclusive, and a second personal conclusion in line with the activity planned for Wednesday 7 December 2016, following the thread of the "urban walk".

8.1-Formal conclusion.

Speculation is part of urbanism, and I think that the author is overly concerned with this oversight. I have followed her development for one thing: I understand that the verb speculate depends on its historical context. In the classical world, the ruling class speculated on the exaltation of its megalomaniac and petulant images. In this sphere, on the other hand, in which the bourgeoisie belongs to the dominant group, it is commercial and investing and therefore invests no longer in its figures, but in a neutral space that aspires to produce profits around its use and exchange value, in addition to its interest in planning a space to suit itself. This is the difference I wanted to make.

It is not the same because of that. Considering that they share that speculative idea, I assume that most of this and other cities have been built through such urban speculation, the interests behind it are distinctive of those initiatives and their projection. I live in a layout that was literally created through the Corte Inglés. And yes, I recognise that this has surely been the driving force that dynamised the neighbourhood, determining its planning and its spatial and social consequences.

Knowing the reality, however sordid and speculative it may be, is necessary. I was unaware of much of the development of Ordoño II and now I understand this space better. Both in the urbanistic and historical sense, as well as in the social consequences it has generated. Just as I now recognise the dynamising value of these initiatives, (more lucrative than social). And I am therefore grateful to have been able to participate in this approach through dynamics that have a great impact on our society, but which should never move in areas where blind spots prevail for the ordinary citizen who then pass through them.

8.2-Personal conclusion: A stroll along Avenida de Ordoño II.

For an afternoon I walked along the avenue of Ordoño II with a notebook in my hand and the information I have collected in the previous pages, relatively fresh. I walked along it and jotted down the personal impressions that came to me, guided above all by the tender urbanistic intuition that I now have.

There is no mercy for the old, unless it is vintage.

Strolling along Avenida de Ordoño II, you come across countless passers-by, all of them busy. Some go into shops, others come out of them, the whole street seems to be immersed in a frenzy of chores. It is not really a street for a leisurely stroll, this abundant traffic has forgotten or is unaware of the old landmark of the "Paseo de las Negrillas". Nobody strolls any more, not even when they are shopping, neither early risers, nor late risers, nor night owls enjoy their steps on Ordoño II.

It was hard not to go with the flow, but I managed to get out of the flow, I even discovered some furtive glances that tried to delve into my anomalous contemplative attitude, it is not an ideal place for observation, (I thought, surrounded by those galloping and determined rhythms), even so I was able to distinguish something.

With my back to the Plaza de Santo Domingo I look at the other side of the street for convenience (I'll go the other way later). I can't help noticing that my pavement seems to be much more heavily used, almost all of them are buildings of more than five floors and made of new materials. I guess they are no more than forty years old. On the other side of the street I can see smaller buildings, three and even two storeys high, nothing like the mastodons that shelter me on this side, projecting a sombre and intimidating effect on me.

I keep looking at these façades to my right and see a decontextualised vigour in them. I also think that the signs that "decorate" the lower floors do not help in this respect, they offend and are very disrespectful to the architecture that shelters them.

One of them (portal nº 5) has a big red ribbon decorating part of the façade, I notice that there is a pharmacy at the bottom and that it is also old. There is no unoccupied ground floor, all of them are being exploited by businesses, and yes, all of them are tertiary, all of them offer a service that does not imply community, that in fact make it impossible to create community in this urban and intensive space.

I continue to move on between bank offices and the prospect of impracticable dwellings. A few metres further on I come across a ground floor under construction, number 8 Ordoño II is being intervened before my eyes, and I ask the workers what they are going to put in that business, to which one of them replies; -A Santander. I understood him, I said goodbye with "as if there weren't enough..." and continued walking. Another bank for the long list of bank branches that have colonised this urban space.

It is becoming increasingly clear to you that you are walking through an open shopping centre. But nobody could live in a shopping centre. A little further on I come across another process under construction, number 14, a three-storey building, which I guess is old because they have forced it to keep the lower façades. Now it has six floors, they are making extensive use of the building land, and they are also planning an interior space in the form of a shopping gallery that connects Ordoño II with the street behind it.

I also notice that in the upper windows of the older, neglected buildings there are a lot of posters stuck up. They always appear on old buildings and on them you can read: "for sale" or almost always "for rent". I wonder if they are still on the Avenida de Ordoño II, and I don't think so, because these buildings belong more to the old Paseo de las Negrillas.

The exchange value and use of these properties is also outdated, nobody invests in old, dysfunctional and anachronistic properties, because they cannot compete with the new ones that have all the contemporary services, nobody has bothered to adapt them, to put a lift in their old-fashioned bowels, they will simply let them die until they meet the same fate as that number 14.

A little further on I come across the first supermarket in the street, and I've already gone halfway down it. I won't see any more; the shop assistants, the clerks, the bankers, the workers, all these people who live off this street during the day will only inhabit it for a few hours, they will soon go home and the street will be empty and deserted after all the day's fatigue. If you pass by once all the shops, banks, offices and administrative offices have closed, the sensation of passing through this dichotomy is terrible.

It is quite depressing in this sense, I find nothing hopeful for its posterity as a living urban element. Almost at the end of the street, in front of the Alcázar de Toledo chamfer, I see a small shop that seems to have an air of other times, it is called "Mercería Ortega", and at least I want to mention that this sewing redoubt still exists. I decide to go in, tell him what I am doing there and if he would be so kind as to answer me a few questions.

She is the heiress of the family business. It was opened by her father in 1949 and, as she said, referring also to other colleagues in the trenches, "We are the ones who hold on". It gives her a living. It is not a franchise and the rest of the new textile businesses on the street are determined to vilify the concept of haberdashery or to eliminate from the collective vocabulary the "textile repairs" of this trade that no longer has a place in our consumer society.

The premises are very small, but the rent, as the manager of the shop confesses to me, "is too high", but it seems to be the price to pay for a small business on the main avenue, whose rent has been increased by the abolition of the old rents. Meanwhile I think sadly that there is no mercy for the old, unless it is "vintage".

He does not live there, and never did because, among other things, he says: "it is an uncomfortable place to live, most of the houses have no parking, access is difficult and there is nothing left there. He has had memories of the haberdashery since he was two years old, when he was already running around his father's shop.

9-BIBLIOGRAPHY

1- BAYÓN, R. ARCE. *La ciudad de León en el siglo XIX. Transformaciones Urbanas Precursoras del plan de ensanche,* León, 2012, p.120.

2- ORDOVÁS, M. J. GONZÁLEZ *Políticas y estrategias urbanas*, Madrid, 2000, pp.121-122.

3- DURANY, M. P. *La calle Ordoño II de León: De calzada real a eje comercial y de servicios,* Salamanca, 1990, p. 21.

Mª del P. DURANY, *La calle Ordoño II de León: De calzada real a eje comercial y de servicios,* Salamanca, 1990, pp. 24-27.

Mª J. GONZÁLEZ ORDOVÁS, *Políticas y estrategias urbanas*, Madrid, 2000, pp. 121-122.

VV. AA., *León, Casco Antiguo y Ensanche. Guía de Arquitectura*, León, 2000, pp. 23-31.

Mª J. GONZÁLEZ ORDOVÁS, *Políticas y estrategias urbanas*, Madrid, 2000, pp. 121-122.

M. SERRANO LASO, *Arquitectura doméstica en León a principios de Siglo (1900-1923). La pervivencia del eclecticismo*, León, 1992, pp. 25-50.

J. R. CABALLERO CHICA, *La arquitectura de la ciudad de León en su fase inicial (1907-1919),* Master's thesis defended at the University of León, León, 2017, pp. 28-30.

J. HERNANDO CARRASCO and M. SERRANO LASO, "Arquitectura contemporánea. Del neoclasicismo a la posmodernidad", in *Historia del Arte en León*, ch.16, León, 1990, pp. 263-264.

S. TOMÉ FERNANDEZ, "La segunda fase de ocupación del Ensanche Leonés: el proceso de renovación desde los años 60", in L. LÓPEZ TRIGAL (Ed.), *Los Ensanches en el urbanismo español, el caso de León,* Madrid, 1999, pp. 115-119.

T. CORTIZO ÁLVAREZ, "El Ensanche de León. Project and first occupation", in *Ibidem,* pp. 104-112.

E J. R. CABALLERO CHICA, *Op.cit.*, pp. 28-30;

T. CORTIZO ÁLVAREZ, *Op.cit.,* pp. 104-112.

S. TOMÉ FERNÁNDEZ, *León, los ríos en el paisaje Urbano,* Gijón, 1997, pp.59-60.

Mª del P. DURANY, *Op. cit.,* pp. 24-38.

E. FERNÁNDEZ GARCÍA, *León y su actividad escénica en la segunda mitad del siglo XIX*, Doctoral thesis defended at the UNED, Madrid, 1997, pp. 70-78.

S. SANTOS VALERA, "La gota de leche en la ciudad de león: una institución benéfica municipal", *Argutorio*, nº10, 2003, pp.27-28.

M. ANDRÉS EGUIBURU*, Op.cit.,* pp. 113-116.

J. C. PONGA MAYO. *El ensanche de la ciudad de León...,* p. 196.

M. ANDRÉS EGUIBURU*, Op.cit.,* pp. 110-113.

Unkown (14 January 2013). El Ensanche de León [Blog entry]. In: Walking BCN. 2012-13. https://caminarbcn12-13t.blogspot.com/2013/01/el-ensanche-de-leon.html

Olaizola Elordi, J. [Juanjo] (18 December 2023). The railway arrives to León [blog entry]. In: Historias del Tren. https://historiastren.blogspot.com/2023/12/el-ferrocarril-llega-leon-i.html

Vergara Pedreira, S. [Susana] (4 April 2021). Those four houses of Ordoño. La Revista de El Diario de León. https://www.diariodeleon.es/monograficos/revista/210404/1251483/cuatro-casas-ordono.html

Printed by Books on Demand GmbH, Norderstedt / Germany